To the Park

Dona Herweck Rice

Publishing Credits

Rachelle Cracchiolo, M.S.Ed., *Publisher*
Conni Medina, M.A.Ed., *Managing Editor*
Nika Fabienke, Ed.D., *Content Director*
Véronique Bos, *Creative Director*
Shaun N. Bernadou, *Art Director*
Valerie Morales, *Associate Editor*
John Leach, *Assistant Editor*
Courtney Roberson, *Senior Graphic Designer*

Image Credits: All images from iStock and/or Shutterstock.

Teacher Created Materials
5301 Oceanus Drive
Huntington Beach, CA 92649-1030
www.tcmpub.com

ISBN 978-1-4938-9838-1

Printed in China
Nordica.082018.CA21800936

When we go to the

 , we will  .

park

slide

When we go to the

park

, we will

swing

.

When we go to the

park

, we will

run

.

When we go to the

 , we will  .

park walk

When we go to the

 , we will  .

park eat

When we go to the

park
, we will

drink
.

When we go to the

park

, we will

climb

.

When we go to the

 , we will  .

park jump

When we go to the

park

, we will

bike

.

When we go to the

 , we will  .

park kick

High-Frequency Words

New Words

go
we
when
will

Review Words

the
to